JANICE VANCLEAVE'S
WILD, WACKY, AND WEIRD
SCIENCE EXPERIMENTS

MANY MORE OF JANICE VANCLEAVE'S
WILD, WACKY, AND WEIRD
ASTRONOMY
EXPERIMENTS

Illustrations by
Jim Carroll

New York

This edition published in 2018 by
The Rosen Publishing Group, Inc.
29 East 21st Street
New York, NY 10010

Library of Congress Cataloging-in-Publication Data

Names: VanCleave, Janice Pratt.
Title: Many more of Janice VanCleave's wild, wacky, and weird astronomy experiments / Janice VanCleave.
Description: New York : Rosen Publishing, 2018. | Series: Janice VanCleave's wild, wacky, and weird science experiments | Audience: Grades 5–8. | Includes bibliographical references and index.
Identifiers: LCCN 2017011982| ISBN 9781499439410 (library bound) | ISBN 9781499439397 (pbk.) | ISBN 9781499439427 (6 pack)
Subjects: LCSH: Astronomy—Experiments—Juvenile literature.
Classification: LCC QB46 .V36385 2018 | DDC 520.78—dc23
LC record available at https://lccn.loc.gov/2017011982

Manufactured in China

Illustrations by Jim Carroll

Experiments first published in *Janice VanCleave's 203 Icy, Freezing, Frosty, Cool & Wild Experiments* by John Wiley & Sons, Inc., copyright © 1999 Janice VanCleave, and *Janice VanCleave's 200 Gooey, Slippery, Slimy, Weird & Fun Experiments* by John Wiley & Sons, Inc., copyright © 1992 Janice VanCleave.

CONTENTS

Introduction 4

Staging 8

Flash! 10

Darkness 12

Stop! 14

Sweaty 16

Tumbler 18

Blinding 20

Long and Short 22

Circular Path 24

Overhead 26

Lower 28

Sun Time 30

Angled 32

How High? 34

Moon Watch 36

Earthshine 38

The Light Side 40

Moonlight 42

Sky Ball 44

Movers 46

Blackout 48

Shadow Parts 50

Finding North 52

Moon Size 54

Glossary 56

For More Information 58

For Further Reading 60

Index 62

INTRODUCTION

Since the beginning of humankind, people have looked to the heavens to try to understand the stars, the planets, and our sun. In modern times, we have rocketed into space to land on the moon. We built the International Space Station to do research in space, and we have engineered powerful telescopes to peer at the far reaches of the universe. Perhaps humans will someday visit Mars.

Astronomy is the study of the planets, the stars, and other bodies in space. The people who decide to work in the field of astronomy have a variety of career paths to choose from. Some scientists study the planets, and others study galaxies. Some astronomers work to learn more about black holes and the universe. Solar scientists focus on our sun, the star that enables life to exist on Earth. All of these people have something in common: they are constantly asking questions to learn even more about space.

This book is a collection of science experiments about astronomy. What is the difference between your local sun time and clock time? Why is space dark? How does the sun's light affect the visibility of Venus? You will discover the answers to these and many other questions by doing the experiments in this book.

HOW TO USE THIS BOOK

You will be rewarded with successful experiments if you read each experiment carefully, follow the steps in order, and do not substitute materials. The following sections are included for all the experiments.

» **PURPOSE:** The basic goals for the experiment.

» **MATERIALS:** A list of supplies you will need. You will experience less frustration and more fun if you gather all the necessary materials for the experiments before you begin. You lose your train of thought when you have to stop and search for supplies.

» **PROCEDURE:** Step-by-step instructions on how to perform the experiment. Follow each step very carefully, never skip steps, and do not add your own. Safety is of the utmost importance, and by reading the experiment before starting, then following the instructions exactly, you can feel confident that no unexpected results will occur. Ask an adult to help you when you are working with anything sharp or hot. If adult supervision is required, it will be noted in the experiment.

» **RESULTS:** An explanation stating exactly what is expected to happen. This is an immediate learning tool. If the expected results are achieved, you will know that you did the experiment correctly. If your results are not the same as described in the experiment, carefully read the instructions and start over from the first step.

» **WHY?** An explanation of why the results were achieved.

INTRODUCTION

THE SCIENTIFIC METHOD

Scientists identify a problem or observe an event. Then they seek solutions or explanations through research and experimentation. By doing the experiments in this book, you will learn to follow experimental steps and make observations. You will also learn many scientific principles that have to do with astronomy.

In the process, the things you see or learn may lead you to new questions. For example, perhaps you have completed the experiment that observes the moon's phases. Now you wonder if your observations would change during different seasons. That's great! All scientists are curious and ask new questions about what they learn. When you design a new experiment, it is a good idea to follow the scientific method.

1. Ask a question.

2. Do some research about your question. What do you already know?

3. Come up with a hypothesis, or a possible answer to your question.

4. Design an experiment to test your hypothesis. Make sure the experiment is repeatable.

5. Collect the data and make observations.

6. Analyze your results.

7. Reach a conclusion. Did your results support your hypothesis?

Many times the experiment leads to more questions and a new experiment.

Always remember that when devising your own science experiment, have a knowledgeable adult review it with you before trying it out. Ask an adult to supervise it as well.

STAGING

PURPOSE To demonstrate rocket staging.

MATERIALS paper cup, 5 oz. (150 ml)
scissors
long balloon, 18 in. (45 cm)
round balloon, 9 in. (23 cm)

PROCEDURE

1. Cut the bottom from the paper cup.

2. Partially inflate the long balloon and pull the open end of the balloon through the top and out the bottom of the cup.

3. Fold the top of the balloon over the edge of the cup to keep the air from escaping as you place the round balloon inside the cup and inflate it.

4. Release the mouth of the round balloon.

RESULTS The attached balloons move forward quickly as the round balloon deflates. The cup falls away and the final balloon speeds forward as it deflates.

WHY? The set of balloons represents a three-stage rocket. Great amounts of fuel are needed to lift and move heavy spacecraft. Each stage of the rocket system has its own set of engines and fuel supply. As each stage uses up its fuel, it drops away, making the rocket system lighter. Each stage lifts the craft until finally the payload is put into orbit or achieves a fast enough speed to leave Earth's atmosphere for a trip into space.

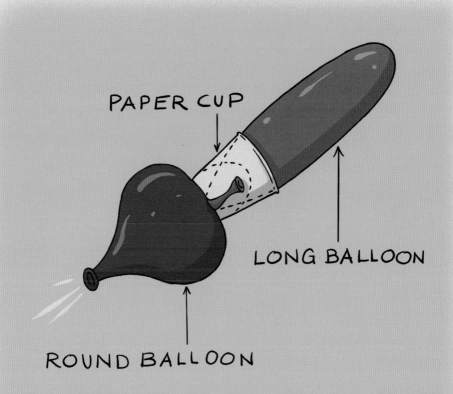

PAPER CUP

LONG BALLOON

ROUND BALLOON

FLASH!

PURPOSE To determine how crystal light might benefit space travel.

MATERIALS wintergreen hard candy
plastic sandwich bag
wooden block
hammer

PROCEDURE

Note: This experiment must be performed in a dark room. A closed closet works well.

1. Place one wintergreen candy in the plastic bag.

2. Place the bag on the wooden block.

3. Position the hammer above the candy.

4. Look directly at the candy piece as you smash it with the hammer.

RESULTS A quick bluish-green flash of light is given off at the moment the candy crushes.

WHY? Crystals broken by pressure give off light. This light is an example of triboluminescence. Crystals such as sugar and quartz give off light flashes when crushed. Crystals that give off light under pressure could possibly be used by engineers in designing the outer shield of space vehicles. It is possible that instruments on Earth could detect crystal light flashes that would indicate trouble spots.

DARKNESS

PURPOSE To demonstrate why space is dark.

MATERIALS flashlight
ruler
table

PROCEDURE

1. Place the flashlight on the edge of a table.

2. Darken the room, leaving only the flashlight on.

3. Look at the beam of light leaving the flashlight and try to follow it across the room.

4. Hold your hand about 12 in. (30 cm) from the end of the flashlight.

RESULTS A circular light pattern forms on your hand, but little or no light is seen between the flashlight and your hand.

WHY? Your hand reflected the light to your eyes, making the beam visible. Space is dark even though the sun's light continuously passes through it because there is nothing to reflect the light to your eyes. Light is seen only when it is reflected from an object to your eyes.

STOP!

PURPOSE To demonstrate how gravity affects inertia.

MATERIALS cereal (your choice)
bowl
milk
spoon

PROCEDURE

1. Pour your favorite cereal into a bowl and add milk.

2. Eat a spoonful of cereal.

3. Raise a second spoonful of cereal to your mouth, but stop before putting the food in your mouth.

4. Observe the position of the spoon and its contents.

RESULTS When the spoon is stopped, the food stays in the spoon.

WHY? This experiment does not present any mystifying results. Of course the food stays in the spoon, but is that always true? No! If you were eating in space and stopped the spoon before it reached your mouth, you would receive a face full of food. Gravity is pulling down with enough force to keep the food from moving forward when the spoon stops moving. Inertia means that an object in motion continues to move until stopped by some force. In space, the inertia of the food would keep it moving after the spoon had been stopped.

Stop!

SWEATY

PURPOSE To determine what happens to water inside a closed area like a space suit.

MATERIALS jar with a lid

PROCEDURE

1. Cover the bottom of the jar with water.

2. Close the lid.

3. Place the jar in direct sunlight for two hours.

RESULTS Moisture collects on the inside of the jar.

WHY? Heat from the sun causes the surface water molecules inside the jar to evaporate (change from a liquid to a gas). When the gas hits the cool surface of the jar, it condenses (changes from a gas to a liquid). Human beings release salty water through the pores of their skin; they perspire. The water from perspiration would evaporate and condense on different parts of the suit, as did the water inside the jar, until the entire inside of the suit was wet and uncomfortable. To prevent this, dry air enters through tubes at one part of the suit, and wet air, along with excess body heat, exits through another tube in a different part of the suit. This circulation of air provides a cool, dry environment inside the extravehicular mobility unit— the space suit.

WATER

TUMBLER

PURPOSE To demonstrate three kinds of satellite movements: roll, pitch, and yaw.

MATERIALS modeling clay

three colored toothpicks: red, blue, and green

index card

marking pen

scissors

PROCEDURE

1. Use modeling clay to form a spacecraft. The length, width, and height of your space vehicle must be less than the length of the toothpicks.

2. Insert the red toothpick through the center of the craft from front to back. This is toothpick A in the diagram.

3. Push the blue toothpick, toothpick B, through the approximate center of the craft from side to side.

4. Stick toothpick C, the green stick, through the spacecraft's approximate center from top to bottom.

5. Draw and cut out a small astronaut from an index card.

6. Stick the paper astronaut into the clay at the top of the spacecraft as indicated in the diagram.

7. With your hands, hold the ends of toothpick A.

8. Roll the stick back and forth between your fingers.

9. Observe the movement of the craft and the other toothpicks.

10. In turn, hold the other two toothpicks rolling them back and forth between your fingers.

11. Again observe the movement of the astronaut and spacecraft as the vehicle rotates.

RESULTS The astronaut and spacecraft rotate around three different axes.

WHY? The three different movements are called roll, pitch, and yaw. Turning around axis A is called a roll. When the craft turns around axis B, the movement is called pitch. Turning around axis C is called yaw. Roll, pitch, and yaw are terms used to describe the motion of a spacecraft. These same terms are also used to describe the movements of airplanes, boats, and cars.

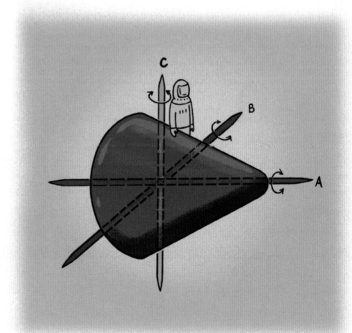

BLINDING

PURPOSE To demonstrate how the sun's light affects the visibility of Venus.

MATERIALS

scissors	white chalk
drawing compass	transparent tape
ruler	pencil
black construction paper	flashlight

PROCEDURE

1. Cut a circle with a diameter of about 2 inches (5 cm) from the black paper.

2. Use the chalk to mark an X about 1/4 inch (0.63 cm) tall anywhere near the edge of the paper circle.

3. Tape the blank back of the paper to the end of the pencil. This will be your model of Venus.

4. Lay the flashlight on a table and turn it on. The flashlight represents the sun. Darken the room.

5. Grasp the end of the pencil and hold it so that the side of the paper with the chalk mark faces you.

6. Hold the pencil at arm's length from your face and about 6 inches (15 cm) in front of the flashlight. Observe the surface of the paper.

7. Move the paper above the light about 4 inches (10 cm) and again observe the surface of the paper.

8. Repeat step 7, moving the paper below the light.

RESULTS When the paper is held in front of the light, the white mark is difficult or impossible to see. When the paper is held above and below the light, the white mark may become more visible, but not much.

WHY? The light behind the paper is so bright that it is difficult to see the surface of the paper. In a similar way, the glare of the sun behind the planet Venus makes it difficult to see Venus. Objects in the sky, such as stars, suns, moons, and planets, are celestial bodies. "Celestial" means "heavenly." Venus is the second planet from the sun, Earth being the third. In its orbit (the path of one body around another) around the sun, as viewed from Earth, Venus is never far from the sun in the sky because it is so close to the sun in relation to Earth. Even when Venus is at its farthest point from the sun, it is difficult to see the planet when the sun and the planet are both above the horizon (an imaginary line where the sky appears to meet Earth). The best time to see Venus, therefore, is before the sun comes up or after the sun goes down.

LONG AND SHORT

PURPOSE To model the relationship between shadow length and the movement of the sun.

MATERIALS scissors
ruler
drinking straw
grape-size piece of modeling clay
flashlight

PROCEDURE

1. Cut one 2-inch (5-cm) piece from the straw.

2. Use the clay to stand the piece of straw on a table. The straw should be perpendicular (at a right angle) to the table.

3. In a darkened room, hold the flashlight on the left side of the straw about 6 inches (15 cm) away. The light should be pointed toward the top of the straw.

4. Slowly move the flashlight directly over the straw, then to the right, in an arc (part of a circle) as shown in the figure. Observe the length of the straw's shadow as the flashlight moves from position 1 to position 3.

RESULTS The shadow of the straw is long when the flashlight is pointing from either side. It shortens as the light moves to an overhead position.

WHY? Earth rotates, which means it turns on its axis (an imaginary line that passes through the center of an object and around which the object rotates). As Earth rotates, the sun appears to rise in the morning and travel in an arc across the sky. In the morning, when the sun is low in the sky, the shadows of objects are long. At noon, the sun is at its highest altitude (angular distance above the horizon) in the sky and shadows are shortest. In the afternoon, shadow lengths increase as the sun's altitude decreases.

Circular Path

PURPOSE To model the movement of the sun's path at Earth's poles.

Materials

5 tablespoons (75 ml) plaster of paris
2 tablespoons (30 ml) tap water
3-ounce (90-ml) paper cup
pencil

white poster board
20-inch (50-cm) piece of string
flashlight
transparent tape

Procedure

CAUTION: Do not wash plaster down the drain. It can clog the drain.

1. Mix the plaster of paris and water together in the cup with the writing end of the pencil.

2. Stand the pencil vertically in the plaster. Allow the plaster to dry. This may take 30 minutes or more.

3. Lay the poster board on a table and set the cup of dry plaster in the center of the poster board.

4. Tie one end of the string to the bulb end of the flashlight and tape the other end to the top of the pencil.

5. In a darkened room, hold the flashlight so that the string is straight and the flashlight is level with the top of the pencil.

6. Move the flashlight in a circle around the pencil, keeping the string straight and the flashlight level with the top of the stick.

7. Observe the direction and length of the pencil's shadow as you move the flashlight.

RESULTS The shadow length remains about the same as it moves in a circular path around the stick.

WHY? The equator is an imaginary line that circles Earth midway between the North and South Poles (northernmost and southernmost points on Earth). When it is summer in the Northern Hemisphere (the region of Earth north of the equator), the sun never sets at the North Pole. Instead, it remains at about the same position relative to the horizon, as it appears to move across the sky. At the same time, it never rises at the South Pole. The same thing happens when it is summer in the Southern Hemisphere (the region of Earth south of the equator). At the poles, the sun's greatest altitude is 23.5° above the horizon. (Keep the cup of plaster for the later experiment titled "Angled.")

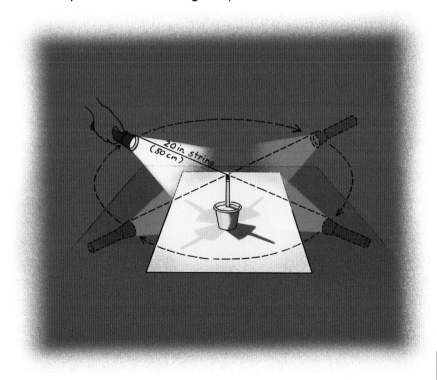

OVERHEAD

PURPOSE To model the highest altitude of the sun.

MATERIALS 6-inch (15-cm) piece of string
protractor with a hole in the center of its base
ruler
two 1-inch (2.6-cm) round orange labels
pen

PROCEDURE

1. Thread one end of the string through the hole in the protractor and tie a knot.

2. About 4 inches (10 cm) from the protractor's base, place the sticky sides of the labels together on the string so that the string is sandwiched between them. Write "Sun" on the labels and draw sunrays.

3. Set the straight edge of the protractor on a table. Holding the protractor in place with one hand, extend the string up to the 90° mark on the protractor with your other hand. Observe the position of the labels.

NOTE: Keep the model to use in the next experiment.

RESULTS A model is made showing the altitude of the sun when it is directly overhead.

WHY? Directly overhead means that the sun is at an altitude of 90°. The sun is only ever directly overhead at or between latitudes (distances measured in degrees north and south of the equator) 23.5°N and 23.5°S.

At summer solstice, the longest day of the year, the sun is directly overhead at latitude 23.5°N, in the Northern Hemisphere (on or about June 21), and at latitude 23.5°S, in the Southern Hemisphere (on or about December 22). During summer solstice in one hemisphere, it is winter solstice, the shortest day of the year, in the opposite hemisphere, and the sun is at its lowest altitude. At vernal (spring) and autumnal equinoxes, on or about March 21 and September 23 respectively, the sun is directly overhead at the equator, latitude 0°. The days are of equal length at the equator all year, but at the equinoxes the days and nights are of equal length all around Earth.

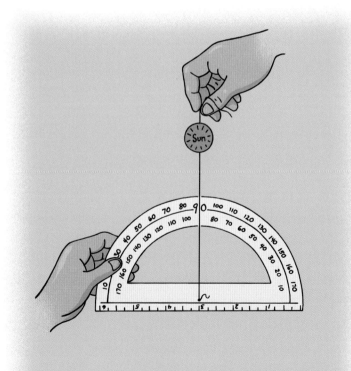

Overhead

PURPOSE To determine the highest altitude the sun reaches at your town's latitude.

MATERIALS model from previous experiment "Overhead"

PROCEDURE

1. Find your town's latitude by looking at a map or checking online with an adult, such as your teacher, librarian, or parent.

2. Determine the angular difference between your town's latitude and latitude 23.5°. For example, at Buffalo, New York, at latitude 43°N, the difference would be $43° - 23.5° = 19.5°$

3. Lay the model on a table with the straight edge facing you. Stretch the string across the 90° mark on the protractor.

4. Move the string the angular distance calculated in step 2 below the 90° mark. In the example for Buffalo, the string is moved 19.5° below 90°. Where does the string cross the protractor?

RESULTS In the example, the string crosses the 70.5° mark on the protractor.

WHY? In the Northern Hemisphere at the summer solstice (June 21), the sun is at its highest altitude, 90° above the Tropic of Cancer (latitude 23.5°N). As one moves away from the Tropic of Cancer, the sun's altitude decreases. For the example in this experiment, the city of Buffalo was calculated to be 19.5° north of the latitude 23.5°N. This means that on June 21, the sun is 19.5° lower in the sky in Buffalo than in a city at latitude

23.5°N. On June 21 at the Tropic of Cancer, the sun is directly overhead, or at an altitude of 90°. On the same date, the sun's altitude decreases as one moves north. In Buffalo, the sun is 19.5° lower than 90°, or at an altitude of 70.5° above the horizon. In the Southern Hemisphere at the summer solstice (December 22), the sun is at its highest altitude of 90° above the Tropic of Capricorn (latitude 23.5°S). As in the Northern Hemisphere, the sun's altitude decreases as one moves away from the Tropic of Capricorn.

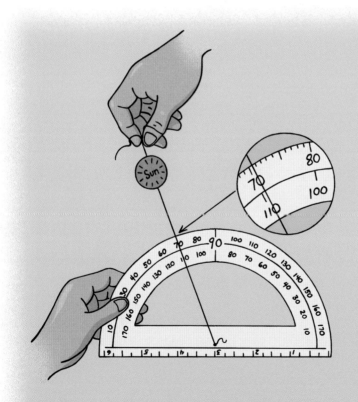

SUN TIME

PURPOSE To determine the difference between your local sun time and clock time.

MATERIALS

pen
ruler
4 x 8-inch (10 x 20-cm)
  piece of corrugated cardboard

watch
directional compass
grape-size ball of clay
short pencil, about 2 inches (5 cm) long

PROCEDURE

1. Use the pen and ruler to draw a 6-inch (15-cm) line down the center of the cardboard. Mark one end of the line N and the other end S.

2. About 40 minutes before 12:00 p.m. (noon) on a sunny day, place the cardboard outdoors on a flat surface. Use the compass to point the N on the line north. *NOTE: Subtract one hour from your watch time if daylight saving time is in effect.*

3. Use the ball of clay to stand the pencil at the S end of the line. The pencil must be perpendicular to the line.

4. Observe the shadow of the pencil as it approaches the line. When the shadow is directly over the line, note the time on your watch.

5. Calculate the difference between the time on your watch and 12:00.

RESULTS The difference between the times will vary depending on where you live.

WHY? Imaginary lines circling Earth from the North Pole to the South Pole are called meridians. These lines are measured in longitude, which is the distance in degrees east and west of the prime meridian running through Greenwich, England, at 0° longitude. In the experiment, the line on the paper represents your local meridian (longitude line). Earth is divided into 24 internationally agreed time zones. Each time zone is about 15° of longitude wide, and local time (clock time) is the same throughout any given time zone. Each zone is centered on a meridian called the time meridian, with about 7.5° of longitude on either side of the meridian. The local time within each time zone is called standard time, but the sun time within each time zone is not the same throughout the zone at any given moment.

This is because as Earth rotates, the sun appears to move from east to west across the time zone. At noon (12 o'clock) local time at locations west of the time meridian, the sun's sky position is generally more to the east, and at locations east of the time meridian, the sun's sky position is generally more to the west.

ANGLED

PURPOSE To make an instrument to measure the altitude of a light source such as the sun.

MATERIALS

yardstick (meterstick)
protractor
masking tape
pencil
3 feet (1 m) of string

cup of plaster from previous
   experiment "Circular Path"
flashlight
ruler
printer paper
helper

PROCEDURE

1. Place the measuring stick on a table. Stand the protractor alongside the measuring stick so the line between the 0° marks lines up with the edge of the stick. Tape the protractor to the stick. The end of the stick opposite the protractor will be called the pointer end.

2. Place a piece of tape across the measuring stick next to the center mark on the protractor. Make a mark across the tape from the center of the protractor. This mark will be called the measuring line.

3. Tear away the paper cup above the hardened plaster and tape one end of the string to the top of the pencil. Set the cup in the middle of the measuring stick.

4. Turn on the flashlight, then darken the room. Hold the cup in place with one hand and hold the flashlight in the other hand over the pointer end about 12 inches (30 cm) from the top of the pencil as shown. Move the cup until the tip of the pencil's shadow touches the measuring line.

5. Ask a helper to extend the string from the top of the pencil to the measuring line, then read the angle where the string crosses the protractor and record it as the altitude of the light.

6. Repeat steps 4 and 5 twice, first raising the flashlight higher, then lowering it. With each experiment, keep the distance between the flashlight and the pencil the same.

NOTE: Keep the instrument for the next experiment.

RESULTS An instrument for measuring the altitude of a light source is made. The angle gets larger or smaller as the height of the light above the measuring stick increases or decreases, respectively.

WHY? The greater the angle, the greater the altitude of the light. The instrument can be used for measuring the altitude of the sun above the horizon. See the next experiment for instructions.

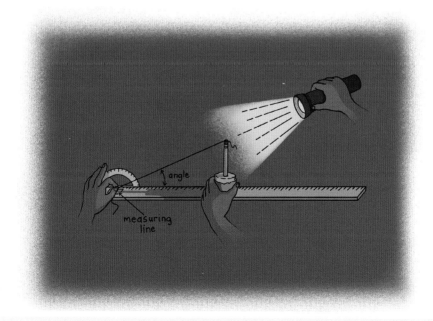

How High?

PURPOSE To measure the altitude of the sun.

MATERIALS

instrument from previous
  experiment, "Angled"
pen or pencil

printer paper
watch
helper

PROCEDURE

CAUTION: Never look directly at the sun. It can damage your eyes.

1. Take your instrument outdoors on a sunny day just before 11:00 a.m.

2. Set the measuring stick on a flat surface with its pointer end facing the horizon directly below the sun.

3. Set the cup in the middle of the stick. Adjust the pointer end of the stick so that the shadow cast by the pencil falls on the stick.

4. Move the cup back and forth on the stick until the end of the shadow touches the measuring line.

5. Hold the cup in place and extend the string from the top of the pencil to the measuring line. Ask a helper to read the angle where the string crosses the protractor.

6. Repeat steps 2 through 5 every 15 minutes until 1:00 p.m. Compare the sizes of the angles.

NOTE: If the shadow is longer than the measuring stick, place two measuring sticks end to end.

RESULTS The angle size increases and then decreases at the same rate.

WHY? The angle measured is equal to the altitude of the sun. In the early morning, the sun is at a low altitude, so the angle is small. As the sun gets higher in the sky, approaching its highest altitude at or near what is called solar noon, the angle increases. After solar noon as the sun gets lower in the sky, the angle decreases.

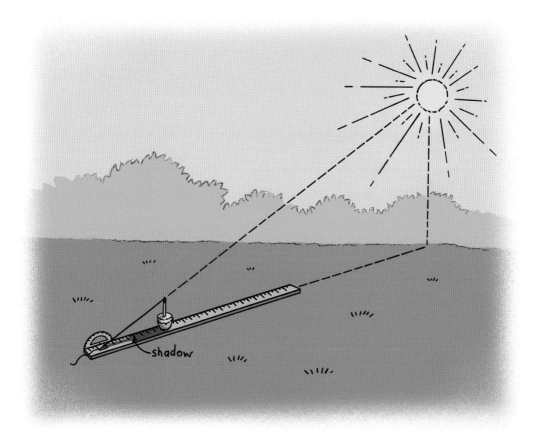

shadow

MOON WATCH

PURPOSE To observe the moon's phases.

MATERIALS sheet of printer paper
pen
ruler
newspaper, calendar, or website with moon phases

PROCEDURE

NOTE: Make no observation for at least three days before and after a new moon—when the side of the moon facing Earth is dark. The new moon is close to the sun and you could damage your eyes if you look at it.

1. Use the paper, pen, and ruler to draw a calendar for 5 weeks.

2. Fill in the dates on the calendar, starting with the day you prepare the calendar. Note that the calendar may include parts of 2 months.

3. Observe the shape of the moon for 29 days. Draw the shape of the moon for each day on the data calendar. For day 1 of observation, check the newspaper, calendar, or website with moon phases listed, then check the Times of Moonrise and Moonset table for observation times for the moon phase for that day.

TIMES OF MOONRISE AND MOONSET

Phase	Moonrise	Moonset
New	dawn	sunset
First Quarter	noon	midnight
Full	sunset	dawn
Third Quarter	midnight	noon

RESULTS It takes about 29 days for the moon to return to the same shape observed on the first night of observation.

WHY? The apparent changes in the moon's shape are called moon phases. Phases are seen because the moon revolves (moves around a center point) around Earth. As it revolves, different amounts of the side of the moon facing Earth are lighted by the sun. The waxing (growing larger) phases as shown are new, crescent, first quarter, gibbous, and full. Following the full moon, the moon goes through the same phases in reverse. These phases are said to be waning, and the quarter phase is called the third quarter. The diagram shows the visible lighted surface in each of the moon's phases.

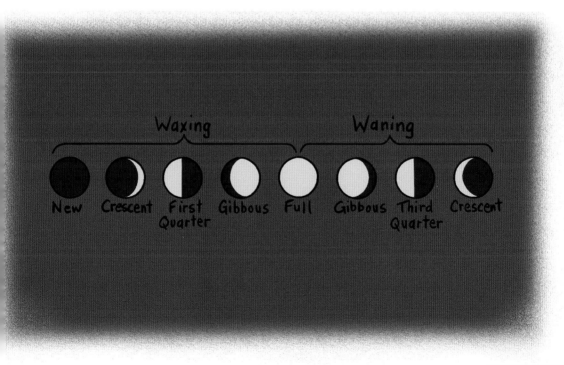

EARTHSHINE

PURPOSE To demonstrate why the faint outline of the entire moon is visible during the crescent moon phase.

MATERIALS

lemon-size piece of blue modeling clay

craft stick

flashlight

3 to 4 books

ruler

cotton ball

PROCEDURE

1. Divide the clay in half and roll it into two balls.

2. Stick a ball of clay on each end of the craft stick.

3. Press one clay piece onto a table to stand the other ball upright. This upper ball will be your moon model.

4. Next to the model, stack the books and lay the flashlight on top so that its bulb is about 2 inches (5 cm) from the clay moon. Turn on the flashlight.

5. In a darkened room, observe the dark side of the clay. Hold the cotton ball about 4 inches (10 cm) away from and to one side of the clay moon, and again observe the dark side of the clay.

RESULTS When you first observed the dark side of the clay moon, it was hard to see any of the surface. When you held the cotton ball near the dark side, the surface of the clay became a bit lighter.

WHY? Light reflects off the cotton ball and illuminates the dark surface of the clay. (The blue color of clay is not significant; any dark-colored surface will be illuminated.) In the same way, sunlight is reflected from Earth's surface onto the moon. This reflected light is called earthshine. Earthshine is happening all the time, but the effect of this reflected light on the moon is more visible during the crescent moon phase. The faint outline of the entire moon is visible during this phase. This phenomenon is sometimes referred to as the "old moon in the new moon's arms."

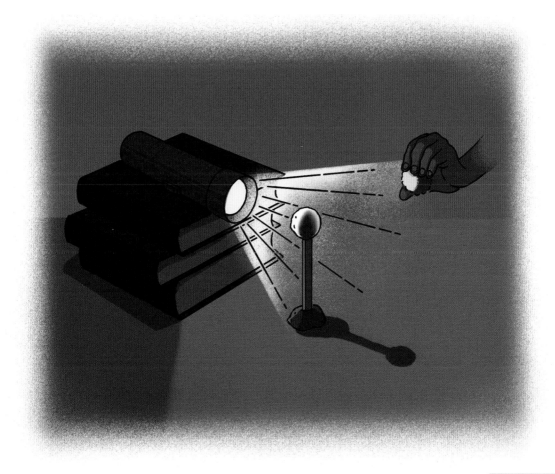

THE LIGHT SIDE

PURPOSE To demonstrate why only one side of the moon is visible from Earth.

MATERIALS drawing compass
sheet of printer paper
pencil
marble-size ball of clay
small paper clip

PROCEDURE

1. Use the compass to draw a large circle on the paper to represent the orbit of the moon. Make a small circle in the center of the large circle to represent Earth.

2. Make a mark on the edge of the moon circle. This will be the starting point of the moon's revolution around Earth.

3. Stick the point of the pencil into the ball of clay, then insert the paper clip in the center of one side of the clay. Stand the pencil next to the starting mark so that the paper clip faces the Earth circle.

4. Slowly move the pencil counterclockwise once along the moon circle, rotating the pencil just enough to keep the paper clip facing the center of the Earth circle. Observe the rotation of the paper clip to determine how many times the clay ball rotates as you move it along the moon circle.

RESULTS The paper clip rotates once, showing that the clay ball makes one rotation on its axis as it moves once along the moon circle.

WHY? The moon makes one complete rotation as it revolves once around Earth. This motion is called synchronous rotation. Because of synchronous rotation, the same side of the moon is always visible from Earth.

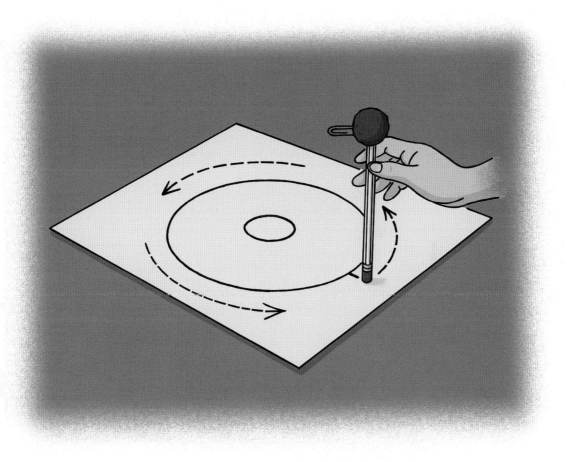

MOONLIGHT

PURPOSE To distinguish between luminous and nonluminous celestial bodies.

MATERIALS

4-inch (10-cm) -square piece of aluminum foil
transparent tape
4-inch (10-cm) piece of string
large shoe box with lid

ruler
scissors
flashlight

PROCEDURE

1. Crumple the aluminum foil into a grape-size ball and tape one end of the string to the aluminum ball.

2. Turn the shoe-box lid upside down, and tape the free end of the string about 2 inches (5 cm) from the corner as shown.

3. Cut a ½-by-2-inch (1.25-by-5-cm) flap about 2 inches (5 cm) from the right corner of one short side of the box. Cut a 1-by-2-inch (2.5-by-5-cm) flap about 1 inch from the opposite corner of the same side.

4. Close the large flap and place the lid on the box so that the ball hangs at the opposite end from the small, open flap.

5. Set the box on a table and look through the open flap toward the ball. Make note of the visibility of the hanging ball.

6. Raise the lid and open the large flap.

7. Repeat steps 5 and 6, shining the flashlight through the open flap toward the hanging ball inside the box.

RESULTS The aluminum ball is not visible or only slightly visible without the light from the flashlight. With the light from the flashlight, the ball appears shiny.

WHY? In this experiment, the ball represents the moon, and the flashlight, the sun. The moon, like the model, is not luminous (giving off its own light). The moon shines mainly because light from the sun, a luminous body, reflects off it. A small amount of the moon's brightness is due to a double reflection. First, sunlight reflects off Earth to the moon, where it is reflected back to Earth. This brightness of the moon is called earthshine.

SKY BALL

PURPOSE To make a model of the celestial sphere.

MATERIALS gummed stars or star stickers

umbrella with 8 sections (preferably a solid, dark color)

PROCEDURE

1. Using the gummed stars to represent the stars of the constellations Cassiopeia and Ursa Major, stick the stars on the inside of the umbrella as shown. The center of the umbrella represents the North Star, Polaris. (Only some of the stars in Ursa Major are used in this experiment. This group of stars is called the Big Dipper.)

2. Hold the umbrella at a slant above your head with the stars of Ursa Major to your left. Face these stars, then turn your head to the right and face the stars of Cassiopeia. Observe the inside surface of the umbrella as you move your head.

RESULTS When you start, you see only the constellation Ursa Major. As you turn your head, Cassiopeia comes into view.

WHY? A constellation such as Ursa Major or Cassiopeia is a group of stars that appear to make a pattern in the sky. Astronomers have designed an imaginary sphere called the celestial sphere to help locate constellations and other celestial bodies. Earth is pictured at the center of this large, hollow sphere with all other celestial bodies stuck on its inside surface. The umbrella in this experiment is a model of the celestial sphere. Turning your head represents Earth's west-to-east rotation. From any given location on Earth, a different part of the sky is seen as Earth rotates. In the

Northern Hemisphere, Earth's axis, represented by the umbrella's center shaft, points toward Polaris, or the North Star. The location in the sky where Earth's axis points is called a celestial pole. Polaris is near the north celestial pole. Thus, it is also called the Pole Star.

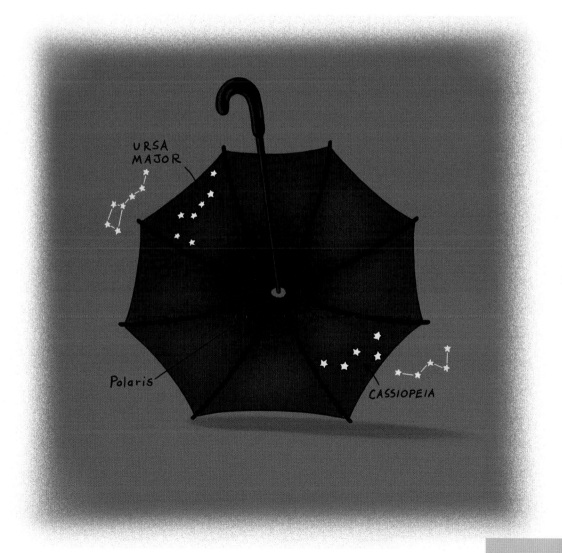

MOVERS

PURPOSE To observe the apparent movement of stars.

MATERIALS

poster board

marker

two bricks, rocks, or other heavy objects

watch

helper

PROCEDURE

1. On a clear moonless night, go outside and find a group of stars that are near the edge of a landmark, such as the roof of your house, a tree, or a telephone pole.

2. Place the poster board on the ground and stand on it. Again locate the same group of stars near the edge of the landmark. Move the poster board if necessary.

3. Ask your helper to use the marker to trace around your feet on the poster board, then to secure the poster board to the ground with the bricks, rocks, or other heavy objects.

4. While standing on the poster board, make a mental note of how close the stars appear to the edge of the landmark.

5. Step off the poster board without moving it.

6. Repeat steps 4 and 5 every hour for three hours or more.

RESULTS The star group moves in relation to your landmark.

WHY? As Earth rotates from west to east, the stars appear to move across the sky. The diagram shows the direction of motion of the star group depending on whether you are facing north, south, east, or west.

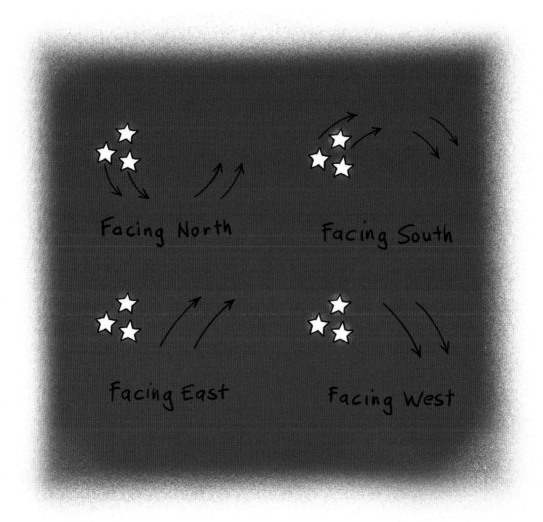

Facing North

Facing South

Facing East

Facing West

BLACKOUT

PURPOSE To demonstrate why the whole Earth is not darkened by a solar eclipse.

MATERIALS

drawing compass
ruler
sheet of white poster board

black marker
lemon-size ball of clay
pencil

PROCEDURE

1. Use the compass to draw a 20-inch (50-cm) diameter circle in the center of the poster board.

2. Use the marker to label the circle Earth.

3. Place the poster board on the ground in a sunny area outside.

4. Stick the clay ball on the eraser end of the pencil. Push the point of the pencil through the center of the circle and into the ground so that the pencil stands upright.

5. Observe the size of the shadow cast by the ball of clay and the amount of the circle that it covers.

NOTE: If the shadow falls outside the circle, push the pencil farther into the ground.

RESULTS The shadow covers only a small part of the circle.

WHY? The passing of one body in front of another, cutting off its light, is called an eclipse. During a solar eclipse, the moon comes between the sun and Earth so that the moon's shadow falls on the surface of Earth. The moon's shadow, like the shadow of the clay ball, covers only a small area. Thus, most of Earth is not darkened during a solar eclipse. For more information about the moon's shadow, see the next experiment.

SHADOW PARTS

PURPOSE To model the parts of a shadow.

MATERIALS

ruler

sharpened pencil

3-inch (7.5-cm) Styrofoam ball

sheet of printer paper

flashlight

PROCEDURE

1. Insert about ½ inch (1.25 cm) of the pointed end of the pencil into the Styrofoam ball.

2. Lay the paper on a table.

3. In a darkened room, hold the ball about 4 inches (10 cm) above the paper.

4. Hold the light about 8 inches (20 cm) above the ball.

5. Observe the ball's shadow on the paper.

RESULTS A dark circular shadow surrounded by a lighter circle is formed on the paper.

WHY? A shadow has two parts: the umbra (darker part of a shadow) and the penumbra (lighter part of a shadow). During a solar eclipse, the moon blocks the sun's light and casts a shadow on Earth. Observers on Earth in the umbra see a total solar eclipse (all of the sun's light is blocked), while those in the penumbra see a partial solar eclipse (part of the sun's light is blocked) and those outside the shadow see no eclipse.

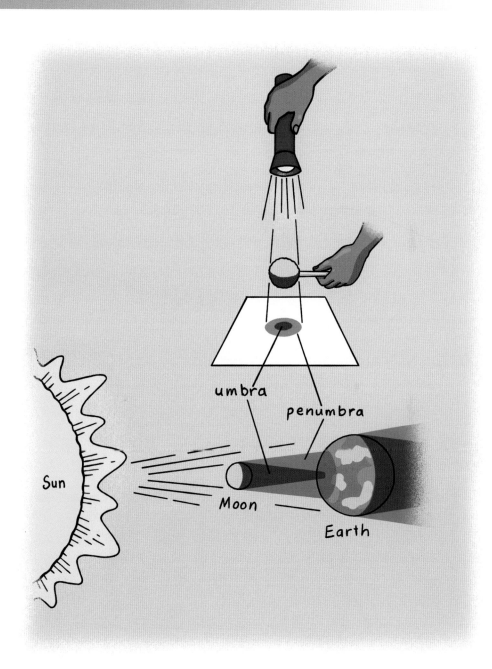

umbra

penumbra

Sun

Moon

Earth

FINDING NORTH

PURPOSE To find compass directions during the daytime.

MATERIALS your shadow

PROCEDURE

1. Stand outdoors on a sunny day at solar noon (when the sun is at its highest altitude and casts the shortest shadows).

2. Face your shadow. You are facing north, and directly behind you is south.

3. Hold your arms out to your sides. Your right hand points to the east, and your left arm points to the west.

RESULTS A sun's shadow is used to find general compass directions.

WHY? In the Northern Hemisphere, shadows point north at solar noon (south, if you are in the Southern Hemisphere). Using shadows at solar noon, you can find the other compass directions. East is to your right, west to your left, and south is behind you.

MOON SIZE

PURPOSE To show how the moon's distance from Earth affects its apparent size.

MATERIALS grape-size ball of clay
two sharpened pencils
3-inch (7.5-cm) Styrofoam ball

PROCEDURE

1. Place the ball of clay on the point of one of the pencils and the Styrofoam ball on the other pencil's point.

2. Hold the pencil with the Styrofoam ball at arm's length so that the ball is in front of your face.

3. Close one eye and hold the pencil with the clay ball so that the ball is in front of but not touching your open eye. Slowly move the clay ball away from your face toward the Styrofoam ball. As you move the clay ball, observe how much of the Styrofoam ball is hidden by the clay ball at different distances.

RESULTS The closer the clay ball is to your face, the more it hides the Styrofoam ball.

WHY? The closer an object is to your eye, the bigger it appears. The small ball of clay can totally block your view of the larger Styrofoam ball. In the same way, the moon, which has a diameter of 2,173 miles (3,476 km), can block your view of the much larger sun, which has a diameter of 870,000 miles (1.39 million km).

During a solar eclipse, when the moon passes directly between the sun and Earth, the moon, the sun, and Earth are in a straight line. In this position, the moon blocks the view of the sun from a viewer on Earth. Even though the moon is about 400 times smaller than the sun, it is about 400 times nearer Earth than is the sun. This makes their sizes look about the same from Earth.

Moon Size

GLOSSARY

ALTITUDE Angular distance above the horizon.

CELESTIAL Heavenly.

CONDENSE To change from a gas to a liquid.

ECLIPSE The passing of one body in front of another, cutting off its light.

EQUATOR Imaginary line that circles Earth midway between the North Pole and South Pole.

EVAPORATE To change from a liquid to a gas.

GRAVITY A force that pulls toward the center of a celestial body, such as Earth.

HORIZON An imaginary line where the sky appears to meet Earth.

LATITUDE Distance measured in degrees north and south of the equator.

LONGITUDE Distance in degrees east and west of the prime meridian.

MERIDIANS Imaginary lines circling Earth from the North Pole to the South Pole.

MOON PHASES The apparent change in the shape of the moon.

NORTHERN HEMISPHERE The region of Earth north of the equator.

ORBIT The path of an object around another body, such as a planet moving around the sun.

PENUMBRA The lighter part of a shadow.

REVOLUTION The movement around a central point, as Earth moves around the sun.

SATELLITE A small body moving around a larger body.

SOUTHERN HEMISPHERE The region of Earth sourth of the equator.

SPHERE A ball shape.

FOR MORE INFORMATION

American Astronomical Society (AAS)
2000 Florida Avenue NW, Suite 300
Washington, DC 20009-1231
(202) 328-2010
Website: http://aas.org

The AAS is the major organization of professional astronomers in North America. Locate an observatory near you, find out about Astronomy Ambassadors, and learn about the latest news in astronomy.

National Aeronautics and Space Administration (NASA) Headquarters
300 E Street SW, Suite 5R30
Washington, DC 20546
(202) 358-0001
Website: http://www.nasa.gov

NASA is the premier organization for all things space! Join the NASA Kids' Club, learn about the International Space Station and historic space missions, view solar system photographs, and learn more about space technology.

National Science Foundation (NSF)
4201 Wilson Boulevard
Arlington, VA 22230
(703) 292-5111
Website: http://www.nsf.gov

The NSF is dedicated to science, engineering, and education. Learn how to be a Citizen Scientist, read about the latest scientific discoveries, and find out about the newest innovations in technology.

Royal Astronomical Society of Canada
203-4920 Dundas Street West
Toronto, ON M9A 1B7
Canada
Website: http://rasc.ca

The Royal Astronomical Society of Canada provides many educational resources, including Ask an Astronomer, observation calendars, photographs, and dates of public astronomy events.

Society for Science and the Public
Student Science
1719 N Street NW
Washington, DC 20036
(800) 552-4412

Website: http://student.societyforscience.org

The Society for Science and the Public presents science resources, such as science news for students, the latest updates on the Intel Science Talent Search and the Intel International Science and Engineering Fair, and information about cool jobs and doing science.

WEBSITES

Because of the changing nature of internet links, Rosen Publishing has developed an online list of websites related to the subject of this book. This site is updated regularly. Please use this link to access the list:

http://www.rosenlinks.com/JVCW/Astro

FOR FURTHER READING

Buckley, James, Jr. *Stars and Galaxies*. New York, NY: DK Publishing, 2017.

Buczynski, Sandy. *Designing a Winning Science Fair Project* (Information Explorer Junior). Ann Arbor, MI: Cherry Lake Publishing, 2014.

Datnow, Claire. *Edwin Hubble: Genius Discoverer of Galaxies* (Genius Scientists and Their Genius Ideas). New York, NY: Enslow Publishing, 2015.

Gagne, Tammy. *Women in Earth and Space Science* (Women in STEM). Minneapolis, MN: Core Library, 2017.

Gardner, Robert. *A Kid's Book of Experiments with Stars* (Surprising Science Experiments). New York, NY: Enslow Publishing, 2016.

Gifford, Clive. *Astronomy, Astronauts, and Space Exploration* (Watch This Space!). New York, NY: Crabtree Publishing, 2016.

Greve, Tom. *Astronomers* (Scientists in the Field). North Mankato, MN: Rourke Educational Media, 2016.

Henneberg, Susan. *Creating Science Fair Projects with Cool New Digital Tools* (Way Beyond PowerPoint: Making 21st Century Presentations). New York, NY: Rosen Central, 2014.

Kawa, Katie. *Freaky Space Stories* (Freaky True Science). New York, NY: Gareth Stevens Publishing, 2016.

Kuskowski, Alex. *Stargazing* (Out of this World). Minneapolis, MN: Super Sandcastle, 2016.

Levy, Joel. *The Universe Explained* (Guide for Curious Minds). New York, NY: Rosen Publishing, 2014.

Riggs, Kate. *Moons* (Across the Universe). Mankato, MN: Creative Education/Creative Paperbacks, 2015.

Rockett, Paul. *70 Thousand Million, Million, Million Stars in Space* (The Big Countdown). Chicago, IL: Capstone Raintree, 2016.

Saucier, C. A. P. *Explore the Cosmos Like Neil DeGrasse Tyson: A Space Science Journey.* Amherst, NY: Prometheus Books, 2015.

Spilsbury, Louise. *Space* (Make and Learn). New York, NY: PowerKids Press, 2015.

INDEX

A

altitude of a light source, measuring, 33
astronomers, types of, 4
astronomy, explanation of, 4
autumnal equinox, 27
axis, explanation of, 22–23

B

Big Dipper, 44

C

Cassiopeia, 44
"celestial," definition of, 21
celestial pole, 45
celestial sphere, 44–45
clock time vs. sun time, 31
compass directions, determining, 52
condensation, 16
constellations, 44–45
crescent moon, 37, 38–39
crystal light, 10
crystals, 10

D

diameter
    of moon, 54
    of sun, 54

E

earthshine, 39, 43
Earth's poles, movement of sun's
    path at, 25
eclipse, 48-49, 50, 55
equator, explanation of, 25
evaporation, 16

F

first quarter moon, 37
full moon, 37

G

gibbous moon, 37
gravity, 14

H

horizon, 21
hypothesis, 7

I

inertia, 14

L

latitude, explanation of, 26
local merdian, 31
longitude, explanation of, 31
luminous celestial bodies, 43

M

meridians, 31
moon
   diameter of, 54
   distance from earth and apparent size, 54–55
   phases of, 36–37
   visibility of only one side, 40–41

N

new moon, 37
nonluminous celestial bodies, 43

P

penumbra, 50
perspiration, 16
pitch, 19
Polaris, 45
Pole Star, 45
prime meridian, 31

R

rocket staging, 8
roll, 19

S

safety, 5
satellite/spacecraft movements, 19

scientific method, 6–7
shadow, parts of, 50
shadow length, the sun and, 22–23
solar eclipse, 48–49, 50, 55
space, why it is dark, 12
space suit, air circulation in, 16
standard time, 31
stars, apparent movement of, 47
summer solstice, 27, 28, 29
sun
   diameter of, 54
   highest altitude of, 26–27, 28–29
   measuring altitude of, 35
sun time vs. clock time, 31
synchronous rotation, 41

T

third quarter moon, 37
time meridian, 31
time zones, 31
tips/advice for experiments, 4–5
triboluminescence, 10
Tropic of Cancer, 28, 29
Tropic of Capricorn, 29

U

umbra, 50
Ursa Major, 44

INDEX

V
Venus, visibility of, 21
vernal/spring equinox, 27

W
waning phases of the moon, 37
waxing phases of the moon, 37
winter solstice, 27

Y
yaw, 19